This Phenomenal Life

This Phenomenal Life

THE AMAZING ways we are connected with our UNIVERSE

MISHA MAYNERICK BLAISE

Guilford, Connecticut

An imprint of Rowman & Littlefield

Distributed by NATIONAL BOOK NETWORK

British Library Cataloguing in Publication Information Available

Library of Congress Cataloging-in-Publication Data

Names: Blaise, Misha Maynerick, 1977-
Title: This phenomenal life : the amazing ways we are connected with our
   universe / Misha Maynerick Blaise.
Description: Guilford, Connecticut : Lyons Press, [2017]
Identifiers: LCCN 2016038193 (print) | LCCN 2016039932 (ebook) | ISBN
   9781493026869 (hardcover : alk. paper) | ISBN 9781493028511 (e-book)
Subjects: LCSH: Biology—Miscellanea. | Life (Biology)—Miscellanea.
Classification: LCC QH349 .B53 2017 (print) | LCC QH349 (ebook) | DDC
   570—dc23
LC record available at https://lccn.loc.gov/2016038193

Printed in Malaysia

♾™ The paper used in this publication meets the minimum requirements of American National Standard for Information Sciences—Permanence of Paper for Printed Library Materials, ANSI/NISO Z39.48-1992.

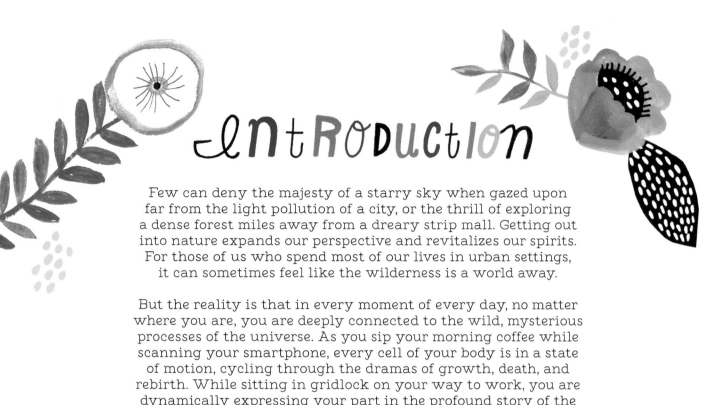

Introduction

Few can deny the majesty of a starry sky when gazed upon far from the light pollution of a city, or the thrill of exploring a dense forest miles away from a dreary strip mall. Getting out into nature expands our perspective and revitalizes our spirits. For those of us who spend most of our lives in urban settings, it can sometimes feel like the wilderness is a world away.

But the reality is that in every moment of every day, no matter where you are, you are deeply connected to the wild, mysterious processes of the universe. As you sip your morning coffee while scanning your smartphone, every cell of your body is in a state of motion, cycling through the dramas of growth, death, and rebirth. While sitting in gridlock on your way to work, you are dynamically expressing your part in the profound story of the connection with all beings in Earth's biosphere. The wilderness isn't something "out there" that we can only access on a camping trip; it is around us and within us at all times. How we build our cities and live our lives is a direct expression of this relationship.

This book is designed as a simple reminder of the staggering ways we are always profoundly connected with the world and all of its creatures, human and non-human alike. It's a reminder that we are one human family sharing one common homeland, and that the Earth is the direct source of all of our material belongings: our clothes, our food, and our cars. Finally, it's a reminder that we are ourselves living embodiments of an infinitely great and mysterious universe.

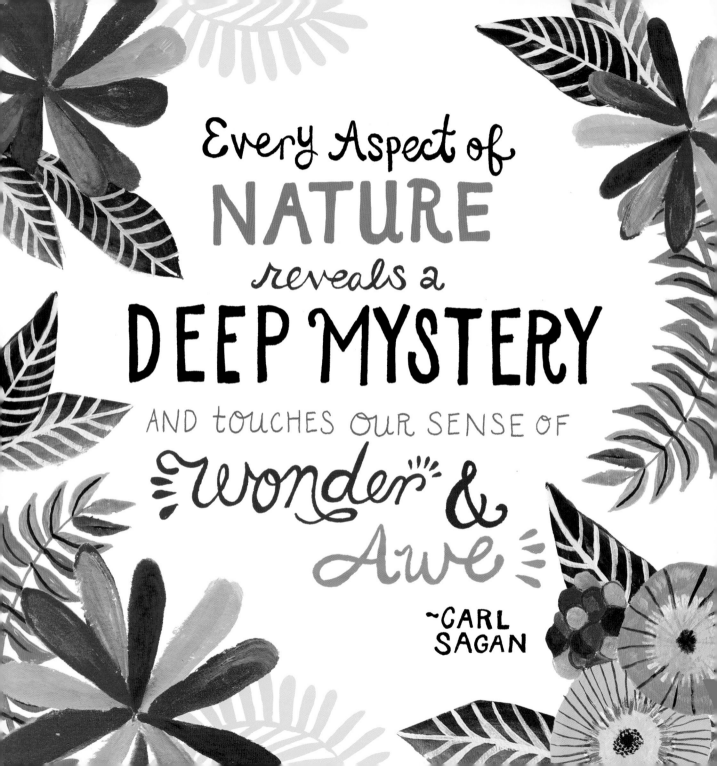

Every Aspect of
NATURE
reveals a
DEEP MYSTERY
AND touches OUR SENSE OF
Wonder & Awe

~CARL SAGAN

Nature is the collective phenomena of the physical world, generally understood to mean everything that is not made by humans.

nature
(jellyfish)

not nature
(can opener)

Maybe you can remember a time when you
experienced the beauty and power of nature:

an UNEXPECTED run-in
with wildlife,

Or maybe
GETTING LOST
in the woods.

Many of us live in cities, and we are seemingly disconnected from the natural world.

BUT, IN FACT,
BY VIRTUE OF LIVING ON EARTH
OUR HOME IS ALWAYS IN NATURE,
NO MATTER WHERE WE LIVE.

Even in a densely populated city, nature can still be found.

In every moment of the day, we directly experience the complex cycles, mysterious processes, and great dramas of the universe, even if we don't realize it. This is the story of the profound interconnection between humans and the universe around us . . .

IN THIS
PHENOMENAL
Life

Almost 96 percent of human body mass is made up of four elements.

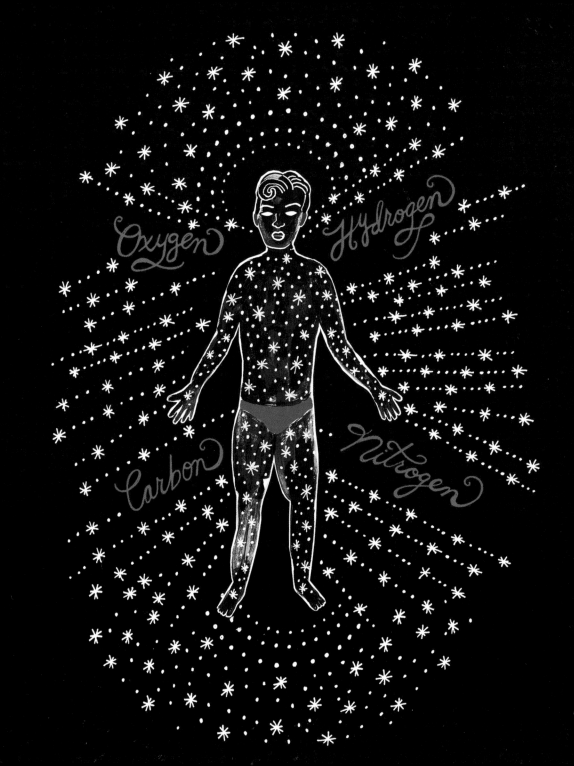

The Sun

73% hydrogen, 25% helium,
2% other

The Earth's Crust

32% iron, 30% oxygen,
15% silicon, 23% other

All of the elements that form life on Earth started off in space. When we gaze upon each pinprick of light in the night sky, we see a distant version of ourselves reflected.

We are literally made from stars, and the atoms inside of us are as ancient as the universe itself.

YOU ARE
THE
UNIVERSE

EXPRESSING ITSELF

AS HUMAN

FOR A LITTLE WHILE.

- ECKHART TOLLE

According to the

BIG BANG THEORY,

all matter in the universe originally was compacted
into an infinitesimally small point called a

Singularity

This point was infinitely dense
and under enormous pressure.

Suddenly, it expanded and moved outward to
create our entire universe. Modern measurements
place this event at about 13.8 billion years ago.

The only elements released by the big bang were hydrogen, helium, and a few of the other lightest elements (like lithium). Time passed (about 100 million years) and as the universe cooled off a bit, these elements began to clump together to form gaseous clouds and eventually stars.

A star is basically an exploding ball of hydrogen and helium, and like a nuclear reactor, it is actively converting this fuel into something else.

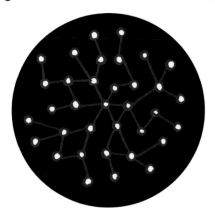

CARBON, NITROGEN, OXYGEN, AND ALL OF THE OTHER ELEMENTS WE ARE MADE OF ARE *forged in the heart of a star.*

Some elements, like

CARBON,

are recycled over and over again incessantly throughout the Earth's biosphere. The amount of carbon on Earth today is the same as when the Earth formed 4.5 billion years ago. Before the carbon atoms in your body became a part of you, they may have cycled through many different forms over millions of years.

One atom of carbon in your body may have once been part of:

A SHELL

A DIAMOND

A VOLCANIC ERUPTION

A FIERCE DINOSAUR

Maybe the same atoms of carbon from the fierce dinosaur that are now within you also have been within other people or things you interact with every day.

In our galaxy alone there are 100 billion stars, and astronomers estimate there are at least 100 billion galaxies in the observable universe. With how many more unknown beings do we share our elemental heritage?

Out of the billions of stars in our universe, the one we probably feel the most affinity with is

The Sun

INTI
THE INCA
SUN GOD

It's the largest object in our solar system and the direct energy source for almost all living beings on Earth.

HELIOS
GREEK Titan
God of the
Sun

About 1 million Earths could fit inside of its colossal gaseous body.

THE MAIN WAY HUMANS GET ENERGY IS BY EATING THE SUN.

We eat the sun by eating plants, or by eating animals that eat plants, or by eating animals that eat other animals that eat plants. Plants are basically intermediaries of the sun's energy. Powered by the energy from the sun, plants combine carbon dioxide and water to make glucose (sugar). A plant captures a ray of sunlight and transforms it into energy-carrying chemicals, which the plant then stores in molecular bonds until you eat it.

ALL OF THE ENERGY
        SURGING THROUGH YOUR BODY
            AT THIS moment
    COMES FROM PLANTS
    that obtained
            that energy from
    THE SUN ITSELF.

Another bit of energy we get directly from sunlight is in the form of

Vitamin D

At least 1,000 different genes governing virtually every tissue in the body appear to be regulated by the way sunlight enters our body in the form of vitamin D.

Even if your house is not strapped up with solar panels, we still power most of our modern cities through the energy of the sun in the form of

fossil fuels.

Coal, oil, and natural gas are formed from the remains of plants and animals that have been compressed under the ground and exposed to the heat and pressure of the Earth's crust for hundreds of millions of years. Plants consumed the sun's energy, animals consumed those plants, and the whole lot of them got smashed together and transformed into our current energy sources.

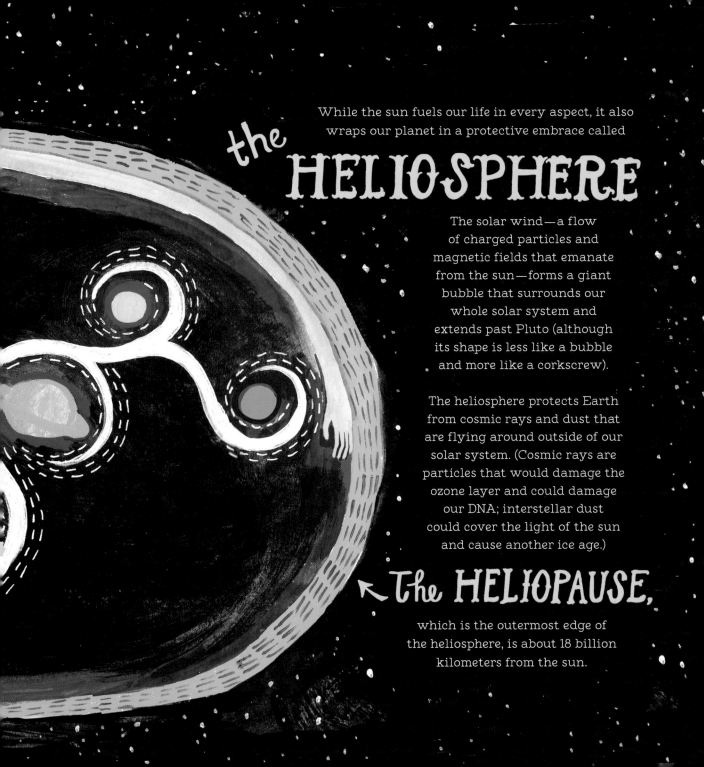

While the sun fuels our life in every aspect, it also wraps our planet in a protective embrace called

the HELIOSPHERE

The solar wind—a flow of charged particles and magnetic fields that emanate from the sun—forms a giant bubble that surrounds our whole solar system and extends past Pluto (although its shape is less like a bubble and more like a corkscrew).

The heliosphere protects Earth from cosmic rays and dust that are flying around outside of our solar system. (Cosmic rays are particles that would damage the ozone layer and could damage our DNA; interstellar dust could cover the light of the sun and cause another ice age.)

←The HELIOPAUSE,

which is the outermost edge of the heliosphere, is about 18 billion kilometers from the sun.

Our solar system is just a tiny part of the Milky Way galaxy, our galactic neighborhood, which is roughly 100,000 light-years across. The Milky Way is part of a larger group of galaxies called

THE VIRGO

SUPERCLUSTER,

which contains about 100 galaxy groups and clusters and is about 110 million light-years in diameter.

In our solar system other planets and moons besides Earth are likely to have water on them. One crucial element needed to harbor life as we know it is liquid water, and our majestic planet is covered in it.

The Earth's surface is about 71 percent water.

WHILE THE UNIVERSE IS INFINITE AND LARGELY UNKNOWN, the Ocean is just as Mysterious

More people have walked on the moon than have explored the deepest underwater regions of our own planet.

While the entire ocean floor has been mapped by satellite from a distance, less than 5 percent of the seabed has been mapped in detail.

Meanwhile, over 90 percent of the Earth's living space is beneath the water. The deepest parts of the ocean, which are cold, dark, and under immense pressure, are home to a diversity of creatures, entirely bizarre to us landlubbers.

ADORABILIS

GIANT ISOPOD

VIPERFISH

DUMBO OCTOPOD

TERRIBLE-CLAW LOBSTER

ANGLERFISH

BLOBFISH

Humans have a ⋛PRIMAL CONNECTION⋚ to WATER

We begin our life in water.
We spend our first nine months
of life immersed in the sea-like
world of our mother's womb.

WE ARE AT ONE WITH **WATER** AT ALL TIMES.

THE HUMAN BODY IS ROUGHLY **65 PERCENT** WATER.

OUR **LUNGS** are **90** PERCENT WATER.

Our brains comprise TISSUE that is almost **80** PERCENT WATER.

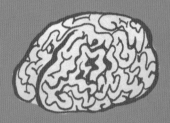

BABIES ARE MOSTLY WATER (**78** PERCENT) when they are born.

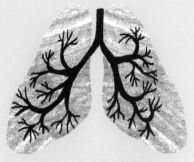

Even our seemingly SOLID BONES are *Flowing* with water.

(31 percent)

EVERYTHING WE CONSUME HAS A
HIDDEN WATER FOOTPRINT:

the amount of water used to grow or power the elements
employed to create a product. Examples of hidden water footprints:

BEEF mmmooooooo

It takes roughly 1,800
gallons of water to produce
one pound of beef.

coffee

About 38 gallons of water are used to make one cup of coffee,

TEA

and about 7 gallons of water are used to make one cup of tea.

EARTH'S WATER MOLECULES
ARE ALWAYS in *movement* AROUND US,
CONSTANTLY CHANGING STATES, FROM
LIQUID *to* VAPOR *to* ICE.

The water you flush down your toilet has just embarked on an epic journey back to the ocean. En route it may be purified and used as someone else's drinking water.

WATER MOLECULES CONSTANTLY CYCLE THROUGH US BY VIRTUE OF WHAT WE EAT.

Here is the percentage water content of some foods we eat:

tomatoes 94%

broccoli 91%

watermelon 92%

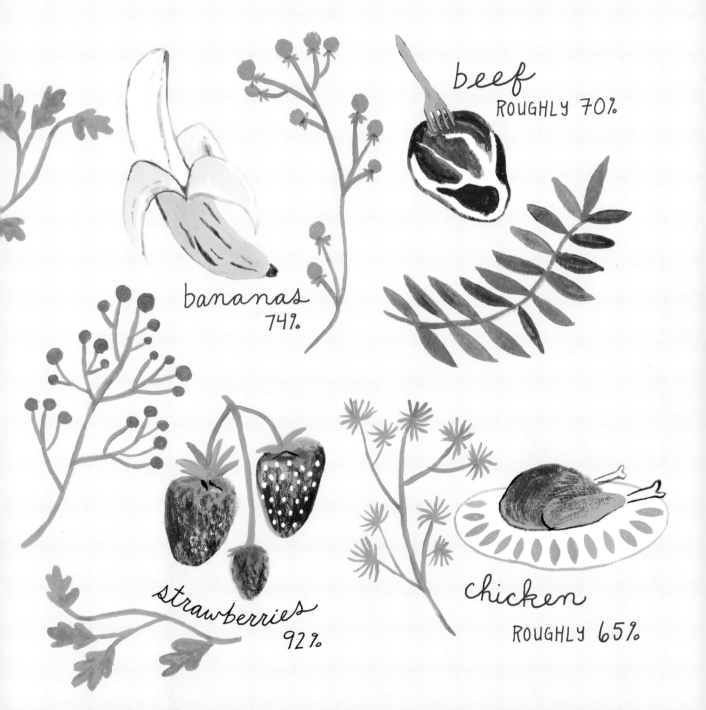

bananas 74%

beef ROUGHLY 70%

strawberries 92%

chicken ROUGHLY 65%

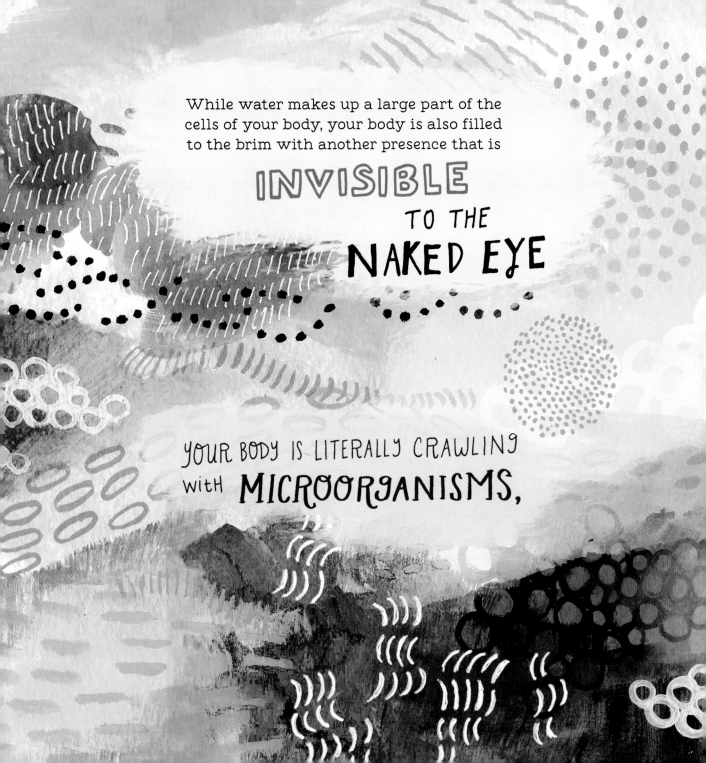

While water makes up a large part of the cells of your body, your body is also filled to the brim with another presence that is

INVISIBLE

TO THE

NAKED EYE

YOUR BODY IS LITERALLY CRAWLING WITH MICROORGANISMS,

so much so that
"YOU"

are largely
made up of
"NOT YOU"
elements.

THERE IS ROUGHLY AS MUCH BACTERIA IN YOUR BODY AS HUMAN CELLS.

For every human gene in your body, there are 360 microbial genes. There are about 10,000 species of microbes (including bacteria, viruses, and fungi) living in you right now.

You are surrounded by your own unique mix of microbes, including bacteria, yeast, and cells, which hover in the air around you in the form of a

"Microbial Cloud."

Microbes are shed from your skin and shoot into the air every time you move or breathe. You exchange microbes any time you are close to another person. Your microbial cloud is so personally distinct it could even be used to identify you.

We potentially leave behind some of our DNA on everything we touch. Just a few cells shed from the outer layer of our skin contain our genetic material.

Microorganisms CONNECT US with our ANCIENT HERITAGE

Microscopic bacteria and microbial archaea (single-celled microorganisms) were the earliest forms of life on Earth.

Humans can trace back what was likely
our shared distant ancestor to a

Microbial Grandmother.

A cluster of microbes found in Western Australia
is one of the oldest forms of life discovered on
Earth; it lived around 3.5 billion years ago. The
microbe group functioned like a small human
society, with each cell hosting individual life
but cooperating with others as a single entity.

The microbe population living in your intestines is called your

MICROBIOTA

(OR GUT FLORA).

It contains tens of trillions of microorganisms, including
at least 1,000 species of known bacteria. Your microbiota is
totally unique to you. Two-thirds of it is completely specific
to your body based on your environment and what you eat.

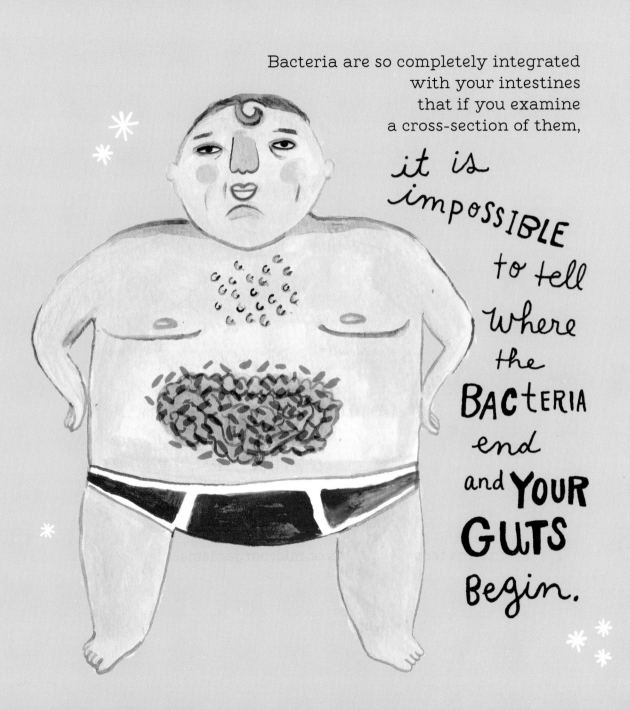

Bacteria are so completely integrated
with your intestines
that if you examine
a cross-section of them,

it is
impossible
to tell
where
the
BACteRIA
end
and YOUR
GUTS
Begin.

In even the cleanest of homes, we are immersed in an invisible world teeming with

Multitudes of Microorganisms

Some 63,000 species of fungi and 116,000 species of bacteria make our homes their homes.

Similar populations of microbes that squat on our toilet seats also snuggle up on our pillows.

ABOUT A **MILLION**

fungi Spores

WILL NESTLE UP WITH YOU TONIGHT WHEN YOU SLEEP.

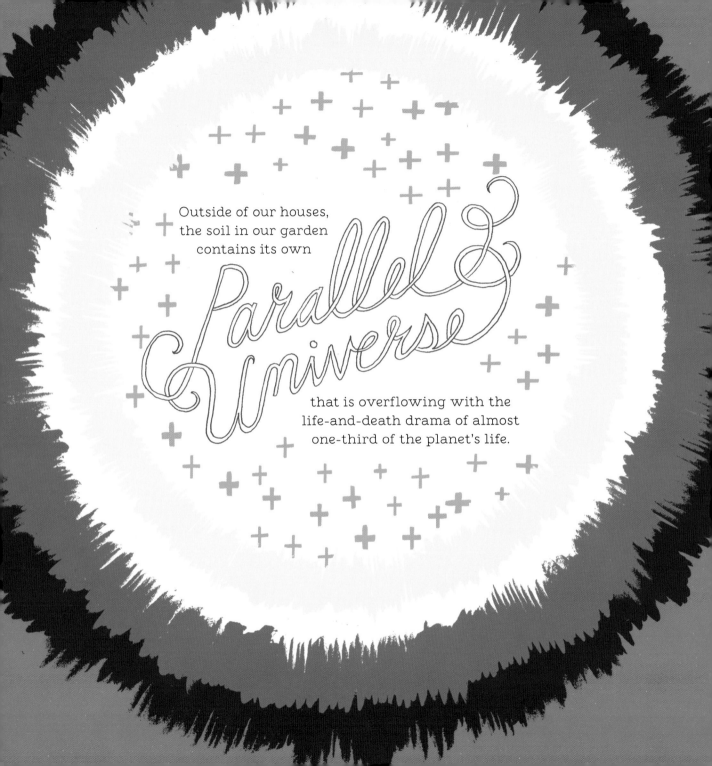

Outside of our houses, the soil in our garden contains its own *Parallel Universe* that is overflowing with the life-and-death drama of almost one-third of the planet's life.

the Soil

FOOD WEB

THIS COMPLEX
ECOSYTEM is so
DENSELY POPULATED
that a
SINGLE SCOOP of SOIL
contains more species of
FUNGI, PROTOZOA, and BACTERIA
than there are plants
and vertebrate animals
in ALL of NORTH
AMERICA.

Within this separate reality, organisms are busily cycling all of the nutrients that support life on Earth

(like nitrogen, carbon, & oxygen).

THESE *denizens* *of the* **dirt**
ARE OUR ALLIES IN
DECOMPOSITION,

breaking down large pieces of organic matter into simpler molecules. This is a crucial ecological process that prevents our communities from being overwhelmed by massive piles of dead stuff.

Digging in the dirt BRINGS US EMOTIONALLY CLOSER to OUR UNDERGROUND COMPANIONS AND ALSO MAKES US FEEL HAPPY.

A specific soil
bacterium called

Mycobacterium vaccae

activates a set of serotonin-releasing neurons
in the brain—the same nerves targeted by many
antidepressants.

While those fungi flourishing in the soil may seem like an alien species,

fungi are actually GENETICALLY CLOSER to HUMANS *than to* PLANTS.

Animals and fungi share a common evolutionary history; their limbs of the genealogical tree branched away from plants perhaps 1.1 billion years ago. We share over half our DNA with fungi. They inhale oxygen and exhale carbon dioxide just like we do.

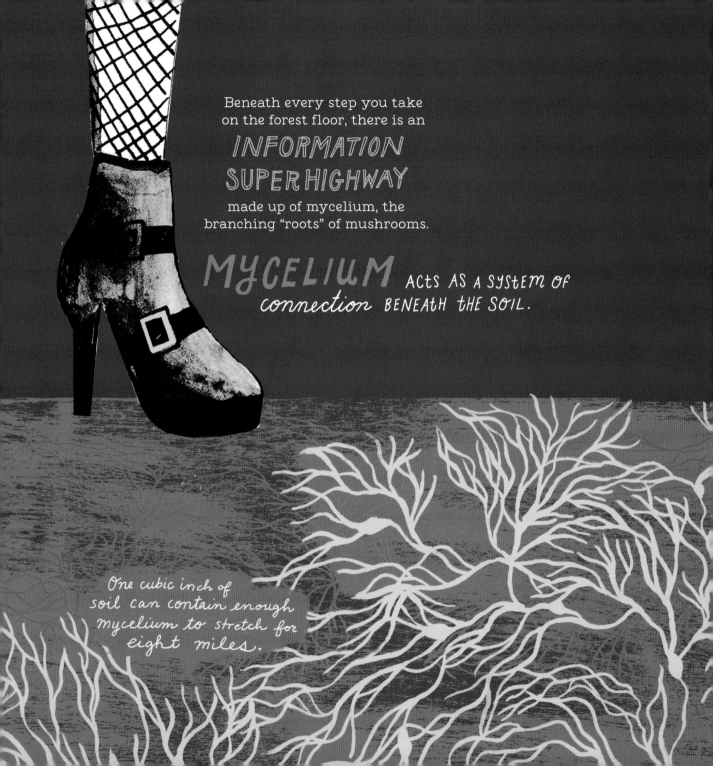

Beneath every step you take on the forest floor, there is an

INFORMATION SUPERHIGHWAY

made up of mycelium, the branching "roots" of mushrooms.

MYCELIUM ACTS AS A SYSTEM OF connection BENEATH THE SOIL.

One cubic inch of soil can contain enough mycelium to stretch for eight miles.

THE GROWTH PATTERN OF
MYCELIUM NETS LOOKS
STRIKINGLY LIKE VISUAL
MODELS OF THE INTERNET.
It closely resembles
the networking of
HUMAN NEURONS,
WHICH ARE ALSO USED TO
transmit information.

Mycelium has been called

"NATURE'S INTERNET."

Under the ground vast networks of mycelium connect different species of plants and trees. They can help plants exchange nutrients, but they also can facilitate

"CYBERATTACKS."

For example, one species of tree is known to emit toxins to prevent the growth of other plants; mycelium helps carry those toxins.

JUST AS you and fungi are veritable cousins, YOU ARE MORE CLOSELY RELATED to YOUR HOUSEPLANT THAN YOU MAY KNOW.

Human life and plant life are incredibly connected. The structure of a chlorophyll molecule is nearly identical to a human hemoglobin molecule (the red pigment found in blood cells).

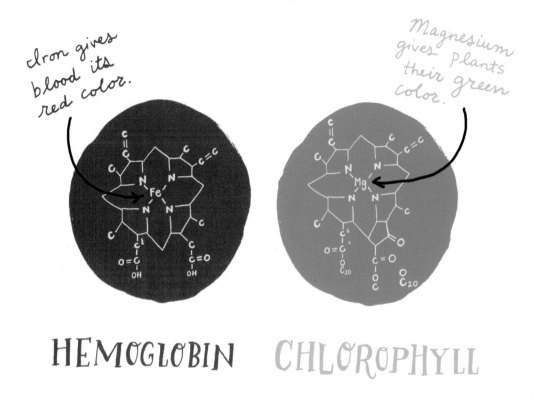

Iron gives blood its red color.

Magnesium gives Plants their green color.

HEMOGLOBIN   CHLOROPHYLL

The only difference is the atom in the middle of each molecule. In plants it is magnesium, whereas in humans it is iron.

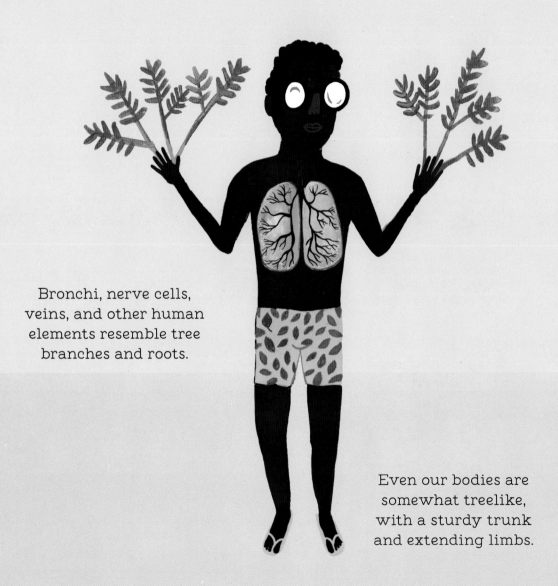

Bronchi, nerve cells, veins, and other human elements resemble tree branches and roots.

Even our bodies are somewhat treelike, with a sturdy trunk and extending limbs.

WE SHARE ABOUT 60 PERCENT OF OUR DNA WITH THE Banana Plant

Plants are aware of the world around them and can sense and react to the world in a way that is familiar to humans.

They can see, hear, feel, taste, and sense gravity and the presence of water, and they can shift direction to avoid obstacles. Plants feel tactile sensations, and some plants (the burr cucumber, for example) are up to ten times more sensitive to touch than we are.

PLANTS HAVE **PHOTORECEPTORS** (cells that detect light) JUST LIKE **HUMANS,** and at the level of perception, plant vision is much more complex than human vision.

Plants can distinguish between red, blue, far-red, and UV light, and they can detect a greater assortment of visible (and invisible) electromagnetic waves. They don't have a nervous system, so they don't see in pictures, but they do translate what they sense visually into cues for growth.

Because plants can track the amount and quality of light and dark in a day, they are able to track the PASSAGE OF TIME.

The system of electric signals produced by plants is similar to the electric signals in human neurons. Plants create neurotransmitters, like dopamine, serotonin, and other chemicals that are also found in the human brain.

PLANT PHOTOSYNTHESIS "CLEANS" OUR AIR BY ABSORBING **CARBON DIOXIDE** AND BY TAKING IN CERTAIN OTHER POLLUTANTS.

NASA studies have shown that houseplants can play a major role in the removal of organic chemicals from indoor air, including benzene, trichloroethylene, and formaldehyde.

Plants take in carbon dioxide from the atmosphere and release oxygen. Humans in turn breathe oxygen and exhale carbon dioxide.

Plant and human BREATHING are thus entwined in a TOTALLY RECIPROCAL RELATIONSHIP.

A RECENT STUDY ESTIMATES THAT THERE ARE ABOUT **3.1 TRILLION TREES ON EARTH;** MORE TREES THAN THERE ARE **STARS** IN THE *Milky Way.*

A leaf on a tree or any aboveground plant, when looked at under a microscope, is covered with tiny breathing tubes called stomata.

STOMATA *means* **MOUTH** *in* **GREEK,**

and like little mouths, these openings pull in carbon dioxide and exhale oxygen.

A square inch of leaf can have hundreds of thousands of these little mouths. In this way plants and trees function as a colossal system of planetary respiration:

Earth's green lungs.

WHILE ABOUT 28 PERCENT OF
EARTH'S OXYGEN COMES FROM
THE RAINFORESTS,
MOST OF IT (70 PERCENT) IS
FROM MARINE PLANTS,
LIKE **PHYTOPLANKTON,**
KELP, AND **ALGAL PLANKTON.**

The bright colors and strong fragrances of plants that bring joy to humans simultaneously attract their allies:

Pollinating INSECTS

The two species are locked in an interdependent embrace that is mutually beneficial.

Scientists have identified about 925,000 species of insects, but as many as 30 million might still be undiscovered.

They are

EVERYWHERE:

Surviving, thriving and festering.

LIKE HUMANS,

INSECTS HAVE A
BRAIN, A NERVOUS SYSTEM,
VISION, AND HEARING.

Amazingly, some insects display intelligence and problem-solving skills that are greater than those of humans, even though their brain is the size of a pinhead.

For example, using just their BODIES, SOIL, and SALIVA TERMITES build some of the most EFFICIENTLY VENTILATED DWELLINGS on Earth.

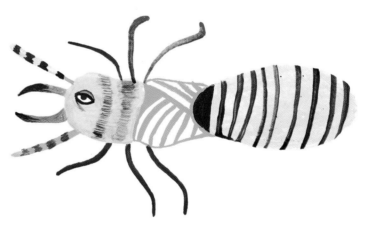

Termite mounds function like a human lung, inhaling and exhaling each day as they are heated and cooled. (Termite mounds can be as tall as 25 feet!)

A bee can find the shortest, most efficient route between the flowers it needs to visit, a mathematical problem that humans have to rely on computer algorithms to solve.

Back at the hive, honeybees use a

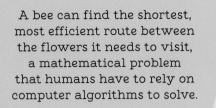

Waggle dance

to then communicate the course to the desired food source. By walking in a figure eight, shaking its abdomen, and fluttering its wings, the bee can broadcast the exact direction and distance to the coveted flower patch, which might be as far as 8 miles (13 kilometers) away.

LIKE HUMANS, BEES HAVE DIFFERENT PERSONALITY TRAITS, AND THEY CHOOSE JOBS ACCORDINGLY.

Honeybee scouts that venture out in search of food express over 1,000 of the genes in their brains differently than other worker bees that wait behind at the hive.

These thrill-seeking bees have increased glutamate activity, which is also correlated with risk taking in humans.

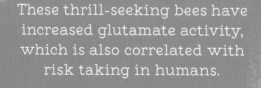

Human life is sustained by bees. They pollinate 70 of the approximately 100 crop species that feed 90 percent of the world. This includes crops that feed cattle, which in turn feed humans.

Incredibly, *humans are not the only creatures who farm.*

Leaf-cutter ants and termites farm fungi.

Some ants have even acquired the skill of *husbandry.*

Herder ants tend to aphids and eat the sugary substance that they excrete (much like we drink the milk of dairy cows). They will even clip the wings of mature aphids so that they cannot fly away. By sharing the skills of farming and husbandry, humans may resemble some insects more closely than any other species on Earth.

Hot on the trail of any insect is a BIRD
waiting to snatch it up for a quick snack. The
sight of a bird in a crowded city parking lot
brings a reminder of wildness to the urban realm.

Human forearms AND bird wings SHARE THE SAME TYPES OF BONES:

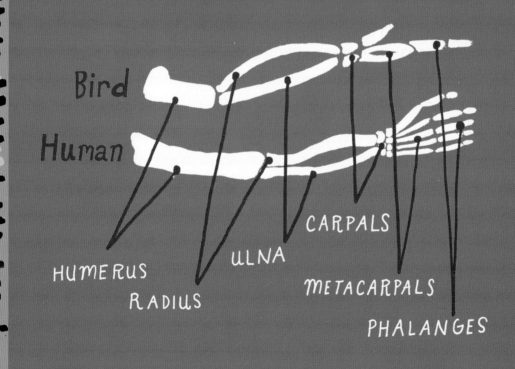

Bird

Human

HUMERUS
RADIUS
ULNA
CARPALS
METACARPALS
PHALANGES

We share about **65 percent of our genes** with **CHICKENS.**

Dozens of the genes involved in the vocal learning that dictates human speech are also active in some songbirds.

Songbirds have specialized vocal learning brain circuits that are similar to those that mediate human speech.

Bird songs reflect the mathematical qualities of human music.

The same simple mathematical ratios that produce harmony that is pleasurable to humans have been found in the North American hermit thrush and other birds. The notes the birds sing fit into the harmonic series, a basic part of human music.

While the major and minor scales these thrush sing are more common in Western music, it has been observed that the hermit thrush also sings pentatonic scales, popular in some non-Western music.

Long BEFORE TEXTing, PEOPLE tRAINED PIGEONS to CARRY MESSAGES BEtWEEN TWO SEt LOCATIONS.

During World War II there was a military unit called the United States Army Pigeon Service, consisting of 3,150 soldiers and 54,000 war pigeons. Over 90 percent of the messages the pigeons transported were successfully received. In 1977 two English hospitals used homing pigeons to transport laboratory specimens. Pigeons have even been used to transport marijuana across cities, or smuggle cocaine into prisons.

INEBRIATED BIRDS slur their words, JUST LIKE HUMANS.

In a research study drunken zebra finches lost their typical song structure and slurred their notes. Strangely, unlike humans, their coordination was not affected by alcohol.

Humans have also partnered with birds of prey to hunt. As early as 1000 BC humans practiced the art of falconry, training birds of prey to hunt quarry (other birds or small animals) and return it to them.

The Art of Falconry

DIFFERENCES BETWEEN
HUMANS AND ANIMALS MUST
INDEED EXIST: MANY ARE ALREADY
KNOWN. BUT, THEY MAY BE
OUTNUMBERED BY SIMILARITIES.
-EDWARD WASSERMAN

We share
84 percent of our
DNA with
DOGS.

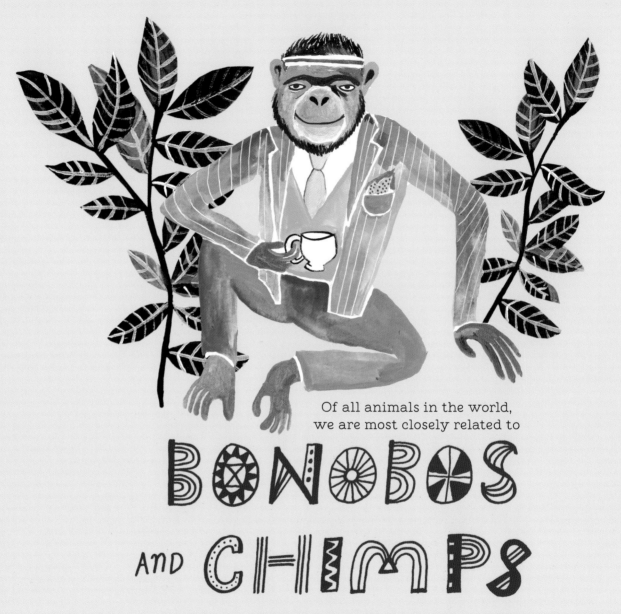

Of all animals in the world,
we are most closely related to

BONOBOS

AND CHIMPS

with which we share about 99 percent of our DNA.

All Mammals have the same ESSENTIAL SLEEP CYCLE and experience THE REM PHASE, the state of sleep characterized by RAPID EYE MOVEMENT, TWITCHING, and Vivid dreams.

(But do animals dream?)

LIKE HUMANS, MANY ANIMALS USE **TOOLS.**

A few examples:

octopuses
(coconut shells to make shelters)

elephants
(branches to swat flies)

ravens
(rocks and sticks for toys)

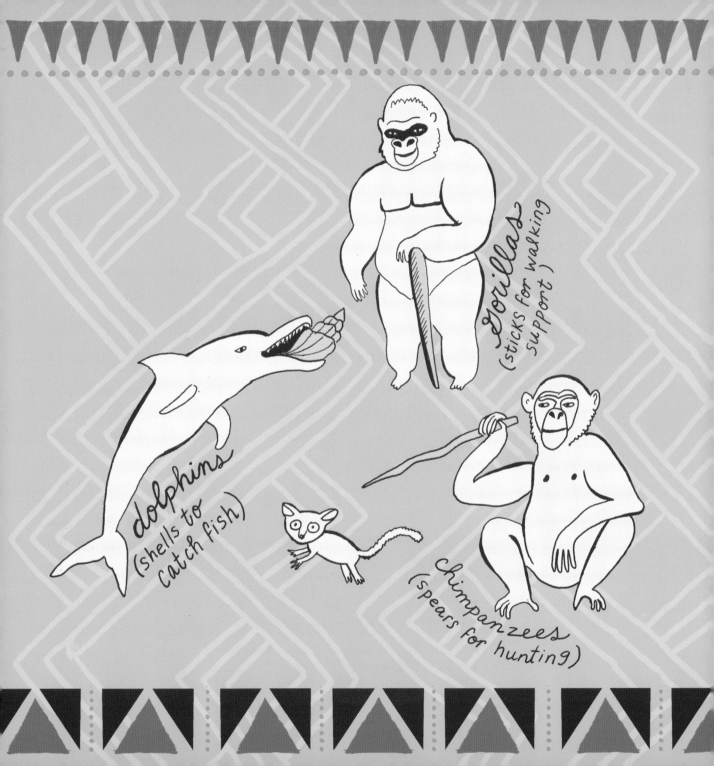

gorillas
(sticks for walking support)

dolphins
(shells to catch fish)

chimpanzees
(spears for hunting)

HUMANS HAVE COLLABORATED with animals FOR CENTURIES to ADVANCE HUMAN CIVILIZATION.

The domestication of animals (and plants) is one of the most critical developments in human history because it provided a stable food source, which in turn contributed to the rise of modern civilization.

A FEW EXAMPLES OF ANIMALS THAT HUMANS HAVE DOMESTICATED AND THE USES FOR EACH:

DOGS

Domesticated at least 12,000 years ago. Uses: meat, herding, protection, hunting companion, etc.

DOMESTIC Turkey

Domesticated at least 1,500 years ago by indigenous peoples of Mexico. Uses: meat, feathers, eggs, and pets.

GOLDFISH

Domesticated in China over 1,000 years ago. Uses: pets, ornamental, and racing.

Silkworms

Domesticated about 5,000 years ago in China. Uses: silk production.

Common Ferrets

Domesticated from the European polecat in 1500 BC. Uses: pets, hunting, racing, and research.

Siamese Fighting Fish

Domesticated in Thailand, 19th century AD. Uses: fighting contests and pets.

Horses

Domesticated across Europe and Asia about 10,000 years ago. Uses: transportation, meat, plowing, warfare, etc.

Sheep

Domesticated in Fertile Crescent of western Iran and Turkey, Syria and Iraq about 10,000 years ago. Uses: meat, milk and milk products, leather, and wool.

PUFFERFISH

BACTERIA

AMOEBAS

CORAL

HUMANS HAVE
SCIENTIFICALLY IDENTIFIED
ROUGHLY **2 MILLION**
SPECIES ON EARTH,
BUT *millions more* LIFE FORMS
ARE BELIEVED TO EXIST,

CACTI

CRABS

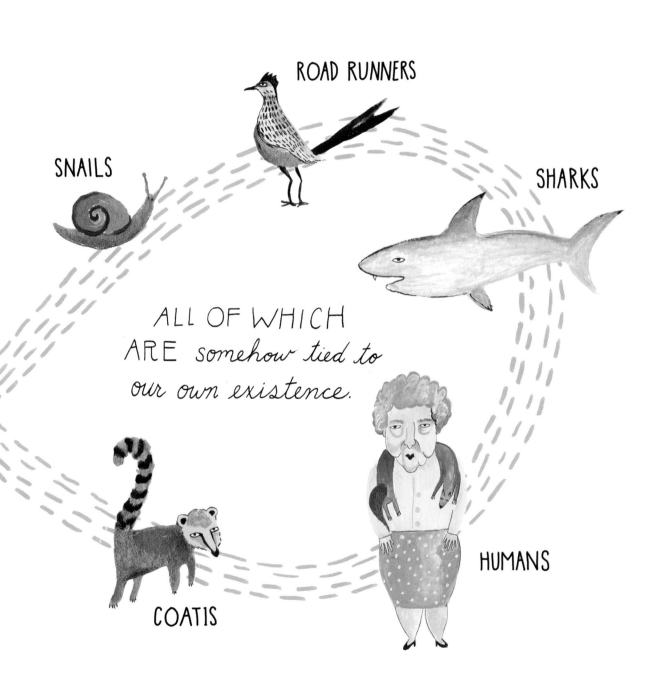

SNAILS

ROAD RUNNERS

SHARKS

ALL OF WHICH ARE somehow tied to our own existence.

COATIS

HUMANS

We don't have to look far to encounter the life forms we are each most closely related to. Despite the great diversity of human culture, ethnicity, and physical traits,

WE ARE ALL, IN FACT, GENETICALLY SPEAKING, 99.9 PERCENT THE SAME.

All persons living right now can trace their family lineage through their mother's side back to a single common female ancestor, known as

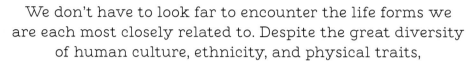

mitochondrial Eve

who lived about 200,000 years ago in Africa, in what is now Angola.

GENETIC SURVEYS SHOW THAT ANATOMICALLY MODERN HUMANS DISPERSED FROM

AFRICA AND SPREAD ACROSS THE PLANET ABOUT 2,000 GENERATIONS (OR ABOUT 50,000 YEARS) AGO.

ALL PEOPLE now on EARTH ARE SOMEHOW RELATED AS DISTANT COUSINS.

The most recent common ancestor of all humans on Earth is estimated to have lived about 3,000 years ago.

People from similar regions share even more recent common ancestors. Every European (not including recent immigrants) shares a common ancestor who lived in Europe only about 600 years ago, around 1400.

The average marriage between people of European extraction is between sixth cousins, who share a common great-great-great-great-great-grandparent.

One in 200 men (about 16 million individuals now alive) is a direct descendant of Genghis Khan, who lived about 800 years ago and ruled the Mongol Empire, the largest empire in history.

GENGHIS KHAN

Approximately 1.5 million Chinese men are direct descendants of Giocangga, the grandfather of the founder of the Qing dynasty, who lived about 500 years ago.

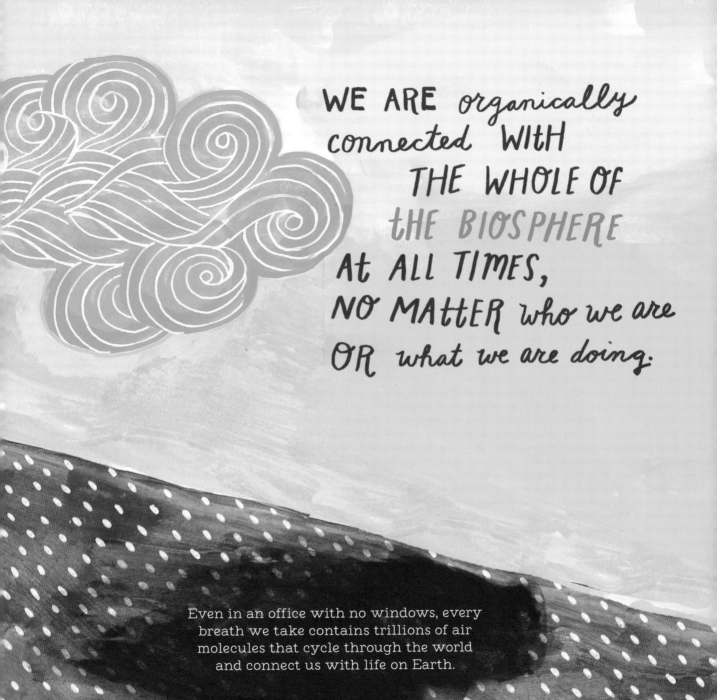

WE ARE *organically* connected WITH THE WHOLE OF THE BIOSPHERE AT ALL TIMES, NO MATTER *who we are* OR *what we are doing.*

Even in an office with no windows, every breath we take contains trillions of air molecules that cycle through the world and connect us with life on Earth.

EVERY moment, THE HUMAN BODY IS ACTIVELY IN A PROCESS OF **REGENERATION**, BASED on **THE ETERNAL CYCLE** of *cell life and death.*

We lose approximately 30,000 cells every hour throughout our lives, and our entire surface layer of skin is replaced about once a year. Some cells take weeks to replace, whereas others take years or even decades.

The "YOU" of 10 years ago may have been made up largely of entirely NEW CELLS compared to the "YOU" of TODAY.

(Our teeth are not living and thus are the only part of the human body that does not regenerate.)

Every atom and molecule within the body is in motion and eventually ends up being excreted, exhaled, or shed. These atoms are replaced directly by elements of the biosphere, in the form of

the food we eat, the water we drink, and the AIR we BREATHE.

Every year, 40,000 tons of COSMIC DUST fall to EARtH.

Eventually all of that debris, which contains oxygen and carbon, iron, nickel, and all the other elements, finds its way into the soil, plants, animals, and our bodies. In this way we have a dynamic relationship with the cosmos above.

SUCH IS THE CYCLE OF

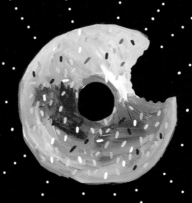

THIS
PHENOMENAL
LIFE.

LOVE REVEALETH WITH unfailing AND Limitless POWER THE MYSTERIES LATENT in the Universe.

THE BAHÁ'Í WRITINGS

thank you

I would like to thank my original agent, Elizabeth Evans, who gave me so much encouragement and praise that I started to believe someone was paying her to do so. Seriously, Elizabeth, thank you for keeping this project alive so that it could see its final incarnation! Thanks to my agent, Laura Biagi, for her meticulous care with every detail of the process, and to my editor, Holly Rubino, who gave me so much freedom and support to bring this wild and unusual project to life!

Special thanks to Felice, Bella, Leila, Farzad, and my sister Monica for your help with critique, editing, and general encouragement along the way. Also, my adorable mom and cousin Karen read through my entire manuscript with great care and interest, which was really heartwarming. Finally, mad love to my husband, Nicholas, for his herculean and unwavering support of my life as an artist. He was always here to help me with my creative process—both writing and illustrating—and he would often take the kids at critical moments so that I could finish my work. He even suggested that I not thank him if it took too much extra time. I love you, you little freak.

A WORD ABOUT SOURCES...

The information presented in this book was compiled from a massive assortment of sources, much of it found online. I was careful to draw from reliable articles citing research from peer-reviewed journals, mostly found in online editions or companions to offline magazines that are established in their field. Websites that I relied on heavily for my research were *National Geographic*, *Scientific American*, the *Smithsonian*, *Discover*, *Science Daily*, *Live Science*, PBS, and *Britannica*. I also found some really useful information on government websites such as the United States Geological Survey and the National Center for Biotechnology Information.

A handful of books served as my inspirational signposts along the way and also provided critical information to ground the topics I explored. These include *Mycelium Running: How Mushrooms Can Help Save the World* by Paul Stamets (2005), *What a Plant Knows: A Field Guide to the Senses* by Daniel Chamovitz (2012), *The Soul of an Octopus: A Surprising Exploration Into the Wonder of Consciousness* by Sy Montgomery (2015), and *The Wild Life of Our Bodies: Predators, Parasites, and Partners That Shape Who We Are Today* by Rob R. Dunn (2011). Rob R. Dunn's online presence at robdunnlab.com includes a seemingly endless supply of online articles and projects exploring the microscopic biodiversity of the planet. His research into the microbial ecology of our households and bodies was enough to make me want to take a shower.

►►►► ABOUT the ◄◄◄◄
AUTHOR
(and illustrator)

Misha Maynerick Blaise is a Canadian-American, raised in the Colorado
Rockies, who now calls Austin, Texas, home. She and her husband are
the co-owners of a green building company and co-creators of two sons.
In her free time, Misha enjoys reading in bed, night swimming, and
watching her husband garden. She also enjoys discussing big ideas over
a strong cup of Persian tea. She can be found at www.mishablaise.com